A project leader is a project manager, but a project manager is not necessarily a project leader.

Published by ╫RICHER Press
An Imprint of Richer Life, LLC

2320 East Baseline Road, Suite 148-214, Phoenix, Arizona 85042
www.richerlifellc.com

Cover Design: RICHER Media USA

Library of Congress Control Number: 2016958665

The Official Leadership Checklist and Diary For Project Management Professionals

Ervin (Earl) Cobb and Jim Grigsby
p.   cm.

1. Project Management   2. Leadership 3. Management
ISBN 978-0-9970831-6-3
(pbk : alk. Paper)

ISBN 13: 978-0-9970831-6-3
ISBN 10: 0-9970831-6-6

Text set is Adobe Garamond

PRINTED IN THE UNITED STATES OF AMERICA

May 2017

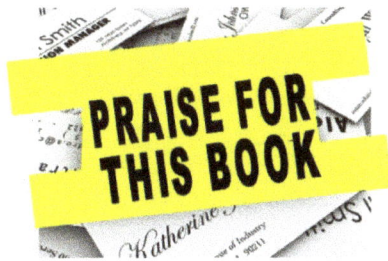

"The book is the consummate guide for PMP professionals to employ and deploy successful project leadership strategies. Most publications focus on the technical aspects of project management while giving little attention to the leadership aspects. This practical guide is a thought provoking resource for project leaders that will enhance project outcomes."

~ Anthony V. Junior, Ph.D., MBA, PMP
Principal, Strategic Consulting Network, LLC

"This book concept to discuss "Leadership" for PM roles is great. This topic is always relevant."

~ Deanna Hawkins, PMP, CEO, 7LampsFX

"Earl Cobb, well-known for helpful, concise business leadership books, has done it again. This time Earl has turned his attention to the world of Project Management, successfully showing PMP's good leadership principles."

~ Doug Russell, PMP, Financial Counselor and Coach

"This book is a high-value quick read, packed with useful information and tips. The advice provided will enhance leadership and project management skills of a seasoned professional or one with the responsibility of a first project."

~ Dr. Kenneth Morton, Infinity Leadership Consulting, LLC

"The book offers a thoughtful, systematic way to ensure that project managers actively lead teams to bring out the best in teams."

~ Yvette DeBois, MD MPH

"Awesome bible for a new PM or a season PM that may need a refresher."
~ Joycelyn Gray, PMP, MBA, CSM

"This book is a quick read for those individuals who wish to share their project management skills and techniques. The book offer opportunities to refresh and reacquaint PMPs with proven concepts in a new and exciting way."
~ Larry McClain, Senior Consultant, CGI Federal

"This book is salient and concise with on point illustrations. It details the difference in skills separating project management from leadership. That leadership is what results in successful project outcomes."
~ Ronald V McBean, Sr.
President, Realm Information Technologies

"The progression of the book from introduction through the seven habits to the diary is very intuitive. This layout will make using the book as a reoccurring reference very easy."
~ Sheldon Johnson
President of Program Management Consulting LLC

"The Official Leadership Checklist and Diary is an excellent resource for leaders at all levels, including project leaders. The book's excellent, highly practical content is simple, straightforward, based on the real world and forever relevant. A great personal, professional and organizational investment!"
~ Dan Nielsen, Retired Healthcare CEO
Founder/CEO/Publisher of America's Healthcare Leaders

"Over my career in Healthcare Financial Management, I have led or managed hundreds of large and complex projects, sometimes with the help of Jim Grigsby. I have always found Jim to be the most organized person on the project. Now I know why! I found the Leadership Checklist and Diary to be right on the mark! I applied some of the principles on a current project as I was reading. The 7 Habits sizzle with common sense. This book has earned a permanent place in my 'toolbox'."

~ John Midolo Managing Partner RCM Strategies

"It reads like an instruction manual that even someone not too familiar with project leadership/planning can understand."

**~ Merlyn Knapp, Principal
Rebound Health Resources LLC**

"This is a great concept for a book. People are still confused about PM. This seems easy to read and easy to implement. This is very well fine. A Welcomed and Much Needed Publication."

**~ D. Anthony Miles, Ph.D., MCP, RBA, CMA, MBC
CEO and Founder at Miles Development Industries Corporation**

"This is an excellent book, front to back. The overall message is not only that project leadership is essential for effective project management, but that leadership must be applied at all stages."

~ Mark M. Cross, P.G., C.Hg., Hydrogeologist / Principal and President, Tucson Operations Manager, Montgomery & Associates

"Unity is strength... when there is teamwork and collaboration, wonderful things can be achieved."

~ Mattie Stepanek

Acknowledgements

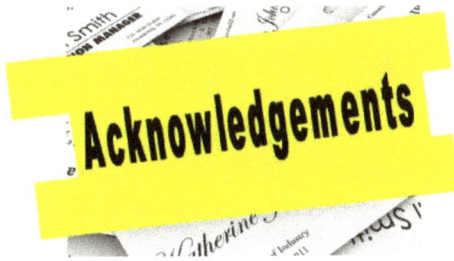

We would like to thank all of those who reviewed this book prior to its release and provided invaluable feedback and input. Without your support, we would not have been able to achieve our goal of offering the millions of Project Management Professionals around the world this unique platform to assist them in significantly enhancing their leadership presence and influence. We would like to offer a special thanks to the following *thought leaders*.

(In alphabetical order)

Ignacio Castro
Retired Director of Quality, ELDEC Corporation

Charlotte D. Grant-Cobb, PhD, MBA

Mark M. Cross, P.G., C.Hg.
Hydrogeologist/Principal and President, Tucson Operations Manager, Montgomery & Associates

Yvette DeBois, MD MPH

Joycelyn Gray, PMP, MBA, CSM

Deanna Hawkins, PMP
CEO, 7LampsFX

Marvin Jackson, MD, FAA
Deputy Regional Flight Surgeon

Sheldon Johnson
President of Program Management Consulting LLC

Acknowledgements

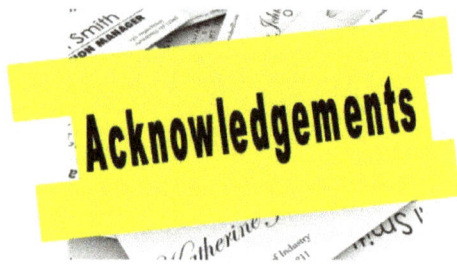

Anthony V. Junior, Ph.D., MBA, PMP
Principal, Strategic Consulting Network, LLC

Merlyn Knapp
Principal Rebound Health Resources LLC

Ronald V McBean, Sr.
President, Realm Information Technologies

Larry McClain
Senior Consultant, CGI Federal

John Midolo
Managing Partner RCM Strategies

D. Anthony Miles, Ph.D., MCP, RBA, CMA, MBC
CEO and Founder at Miles Development Industries Corporation

Dr. Kenneth Morton
Infinity Leadership Consulting, LLC

Dan Nielsen
Retired Healthcare CEO
Founder/CEO/Publisher of America's Healthcare Leaders

Doug Russell, PMP
Financial Counselor and Coach

CONTENTS

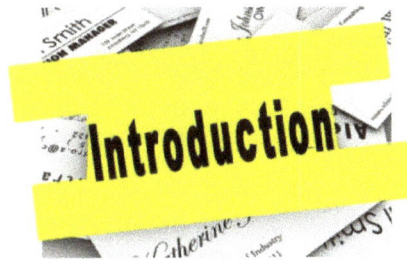

"Leadership is intangible, and therefore no weapon ever designed can replace it." ~**Omar N. Bradley**

Effective Leadership: The Secret Weapon of Successful Project Management Professionals

Projects are temporary endeavors undertaken to create a unique product offering, service or result. From the construction industry to high technology - banking to education - healthcare to customer service and Main Street to Wall Street, millions of critical tasks are accomplished each day by successfully completing projects.

According to the Project Management Institute (PMI), there are over 16.5 million people around the world who diligently work as project management professionals. We believe that most of them would define the key to job success as completing their projects on-time, on-budget and accomplishing all of the original objectives.

The PMI along with the IT Service Company, Gantthead, claim to have relationships with more than 1.2 million of the 16.5 million program management professionals. They indicate that 66.8% of PMI members and 53.9% of Gantthead members are based in North America. They are among many organizations, globally, which provide project management training, software, tools and certifications as a way of enhancing an individual's project management skills and maintaining a standard platform for achieving project success. Many employers use certifications as a means of evaluating a project manager's qualifications and those new to project management are forced to determine which qualifications they should seek in order to gain the best career opportunities.

However, many experienced project managers have learned that managing a successful project involves more than schedules, templates and software. It requires the application of strong interpersonal management skills to work effectively with people in a variety of roles. The most successful project managers are also successful "*project leaders.*"

For some time now, based on our own decades of professional experience in both project and senior management roles along with discussions with dozens of project management professionals, we, not surprisingly, have concluded the following:

"Strong project management skills combined with effective leadership strategies can help you improve your interpersonal communications and to become more influential within your organization. This combination can also help you more effectively guide your project teams through change, deal with conflict and achieve challenging milestones during the entire project management process."

As project management professionals, lecturers and consultants, we have integrated this message into our work as well as our one-on-one engagements with clients, employers and other project management professionals.

However, as published authors who have written extensively about leadership and dealing with daily crises, we have contemplated how we could succinctly document and widely share with others the strong correlation that exists between *"effective leadership"* and *"successful project management."*

We both agreed that if we were to take on such a task, we would have to find a way to package our message into a vibrant, stimulating book – built on a unique platform.

Our goal would be for the book to become a treasured companion to millions of project management professionals around the world and regarded as the perfect leadership *thought guide and performance tracker.*

In addition, the project specific *actions* and *results* recorded in the guide would then serve as a "diary" to be revisited for future projects and/or shared with others to help them enhance their project leadership success.

Somewhat to our own amazement, with the help and collaboration with many colleagues, we were able to create such a self-development platform and publish *The Official Leadership Checklist and Diary for Project Management Professionals.* We have designed this book to be a unique and personal *leadership enhancing complement* to the preeminent global standard for project management as set forward in the *Project Management Body of Knowledge (PMBOK®).*

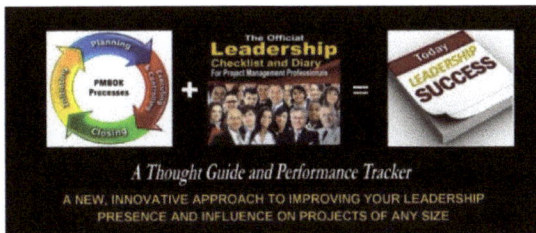

A Thought Guide and Performance Tracker

A NEW, INNOVATIVE APPROACH TO IMPROVING YOUR LEADERSHIP PRESENCE AND INFLUENCE ON PROJECTS OF ANY SIZE

We believe the book is intuitive and self-guiding. However, we do suggest that you, first, read the *"How to Benefit Most from this Book"* section and then...go for it. We are certain that you will find the combination of your project management skills and your heightened leadership competence to be your *secret weapon* for continuous project management success.

"Experience is a safe guide."

~William Penn

How to Benefit Most From this Book

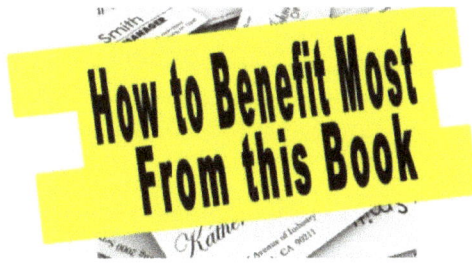

ig Ziglar's advice is as accurate today as when he first coined it; "People often say that motivation doesn't last. Well, neither does bathing - that's why we recommend it daily." We recommend that you consult and benefit from this book daily – utilizing the appropriate *Checklist*, responding to the applicable questions in the *Diary* and using the *Leadership Checklist* as a trusted guide.

Think of *The Official Leadership Checklist and Diary* as your personal brain catalyst – a handy guide containing thought provoking reminders to keep the leadership aspects of your project on track. Such a catalyst can help you lessen the "Oh my God, I forgot" and "This is urgent" moments.

All major projects are complex undertakings. No one can remember or foresee everything. However, as the project leader, you are expected to remember, foresee and be prepared for everything.

Here is an illustration of how you may benefit:

As the project ramps up, you refer to the Project Launch Communication Checklist and stop when you read:

"I have a strategy by which I will communicate clarity of objectives, targets and benefits with each project stakeholder to highlight the merits of the project to their organization."

15

Immediately following is a self-conversation: "Did I do that?" "No!" "I need to take care of that today."

The time you have just invested in consulting the Leadership Checklist and Diary is immediately rewarded. You discovered a forgotten strategic action and quickly formulated a plan to address it.

A few moments or days later, you flip back to the Project Launch section of the Leadership Diary. You see that it is a good time to reflect upon and respond to the appropriate questions regarding "What you will be doing during this project phase to maximize your leadership presence?"

Again, you leave your consultation with the Leadership Checklist and Diary with the confidence that the leadership aspects of your project are on track.

You will be pleasantly surprised that only a few minutes each day – probably not as long as your shower – will amazingly keep the leadership aspects of your projects on track. You will feel more confident because you will be more confident. Your project team, your client/sponsor and your stakeholders will all notice your confidence, your leadership presence and your management strength.

The Leadership *Checklist* is not just another "questionnaire." A questionnaire is simply a list of questions. A list of questions, even within a good book, might be browsed through once for basic understanding; but it will not spur you to revisit it daily and regard it as an insightful, trusted guide – providing insight, vision and direction.

The Leadership *Checklist* contains reassuring "affirmations" of specific actions that should be taken during project performance to increase your leadership presence and influence. Consulting the Leadership *Checklist* frequently can help keep you focused and mentally locked in on the leadership aspects of your project.

A "checked" box indicates that "you" have taken the action. Thus – "I have a plan...I have chosen... or I have decided." This *trail of actions taken*, serves as a distinctive, chronological roadmap and a confirmation of your proactive tendencies and leadership focus. An "unchecked" box indicates that there are still actions to be taken – and later affirmed – during a specific phase of the project.

The Leadership *Diary* uniquely complements the Leadership *Checklist*. The Leadership *Diary* plays a vital role in the process of strengthening your presence and influence on a particular project.

We mutually agree that gaining wisdom from an experience requires reflection. Reflection requires a type of introspection that is essential for any leader. The Leadership *Diary* is structured to be your personal reflective journal. Your thoughtful responses to each of the six questions during each of the five project phases can help you keep focus on the project's leadership priorities, assess your ongoing leadership performance and enhance your overall project planning. At the end of the project, you will own a "private reflective journal" to reference later for further development and/or share with other project managers to help them enhance and grow their professional performance.

We strongly believe that *The Official Leadership Checklist and Diary* will become one of the most valued and trusted tools in your project management toolkit.

"Successful people are simply those with successful habits."

~Brian Tracy

PART ONE

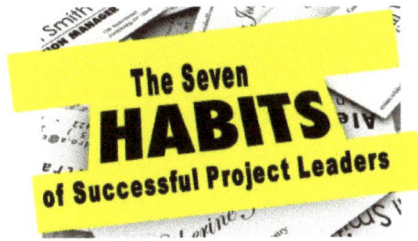

The Seven HABITS of Successful Project Leaders

habit is something we do regularly without consciously thinking much about it. It is an automatic mental and behavioral activity. Habits make it possible for us to do things without spending exorbitant mental effort. In general, good habits are the basis of all professional success. Bad habits usually contribute greatly to professional and personal missteps and downfalls.

For project management professionals, the ability to reliably respond at a rapid pace to multiple project tasks and the associated pressures and relationships is critical to project success. Being able to instinctively provide the correct response, at the correct time and with the correct level of emotional intelligence is what separates *good project leaders* from *outstanding project leaders*.

Likewise, as a successful project management professional, it is imperative to develop more effective project "leadership" skills and habits to lead more complex and more challenging projects.

Based on our experience and research, the following are seven habits and leadership behaviors that we believe can significantly elevate your "game" as a project leader. We are confident that, if adopted, they will assist in placing you in the category of being an *outstanding project leader.*

19

"The art of communication is the language of leadership."

~James Humes

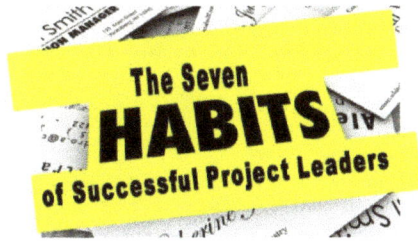

#1 Communicate Vertically, Horizontally and Often

As a successful project manager, you know that you must use communication skills first, and the project tools second. You also know that if you are not making a strong effort to effectively communicate with your team, you will more than likely fail. However, do you realize that simply having effective communications with your team members is only about a quarter of the job?

❑ Project management professionals who are successful on a continual basis know that along with making a strong effort to effectively communicate with their team, they must also work just as hard at effectively communicating with their superiors, their management peers and all other stakeholders (customers, clients, vendors, etc.).

❑ Failure to communicate effectively vertically, horizontally and often, can lead to conflict, create uncertainty and allow the spread of rumors, skepticism and unhealthy relationships.

❑ Do not worry about communicating too much. It has been proven that when it comes to leadership, too much communications is better than too little.

21

❏ If the team members complain about too much information, explain that you are communicating across a wide spectrum and they are free to filter information, as long as they recognize it is their choice.

❏ If the client/sponsor complains, work to develop a communication schedule and format. Document the adjustment and ask them to approve the plan.

❏ Remember, your *thoughts become your actions*. When you make it a habit to monitor and get feedback on the quantity and effectiveness of your communications ...at all levels...and...from all directions...you will always be in a better position to take the actions necessary to stay on course and avoid unnecessary pitfalls.

❏ As a habit, successful project leaders communicate continuously, concisely, compassionately and often.

From a Brighter
Perspective

Professional Football is an Excellent Example of Communicating Vertically, Horizontally and Often

"Excuse me, Coach - But are we the hugs or the kisses?"

The offensive coordinator calls the play into the quarterback's helmet; the quarterback tells the team which play they will run and from what formation. When the team reaches the line of scrimmage, the QB scans the defense and uses signs (or fake signs) and verbal signals to his players; at the same time, the center is telling the other linemen which blocking scheme they will employ. Then, they run the play. All of this occurs in 35 seconds or less.

After the play is completed, the offense repeats the same actions until they score or turn the ball over to the other team.

When the QB is on the sidelines, he communicates with the offensive coordinator and the head coach, while reviewing snapshots of the defensive formations on a tablet. Constant communications at all levels of the organization produces "wins."

"A leader is one who knows the way, goes the way, and shows the way."

~ John Maxwell

The Seven HABITS of Successful Project Leaders

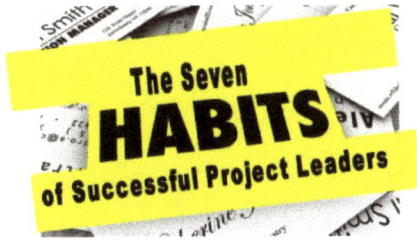

#2 Plan as a Team, Execute as a Team, Win as a Team

To have a great project team, there is no magic recipe for success. Our experience indicates that the combination of strong leadership, effective communication and access to adequate resources contribute to productive collaboration. However, it all comes down to having team members who understand each other and work well together to achieve ultimate success. Having a project leader who has made it a habit of consistently fostering the right mix of planning, trust, ambition and encouragement among his or her team members is crucial to the execution and success of any type and size project.

❑ As set forth in the Project Management Body of Knowledge (PMBOK®), your success as a Project Management Professional will depend on your skills, ability and knowledge in planning.

❑ However, since every project is unique in the problems that arise, the priorities and the environment in which it operates, successful project leaders realize that not involving the project team in the planning process, as much as possible, can lead to critical knowledge gaps as to the true project path.

❑ As we all know, a good plan is dynamic, flexible and constructed to capture and track the actual project path as well as crystallizing the path in the minds of all team members.

❑ The key to fostering the right mix of planning, trust, ambition and encouragement is cultivating a sense of ownership.

❑ By including the entire project team in the planning process, they will gain a personal sense of responsibility and ownership for the project and their individual tasks.

❑ If team members do not feel a personal sense of responsibility and ownership for the project, they will distance themselves from it at the first sign of trouble.

❑ When all aspects of the project execution are recognized as valuable contributions to the team's overall success, all of your team members will feel that they are working together toward a common goal. As a result, when the project successfully crosses the finish line...it will be seen and will feel like a team win.

❑ As a habit, successful project leaders not only value the idea of planning, executing and winning as a team, they also understand that it is their role to create a project environment which facilitates and encourages the maximum level of team involvement.

From a Brighter
Perspective

Successful Projects are the Result of Teamwork
from the Initial Plan to the Exit Meeting

**"Yes. There is room for the unicorn, unless the EPA
requires larger recreation areas for the elephants."**

*God gave Noah a plan for the ark and Noah used a team of laborers to
build it to God's specifications. Throughout the process, Noah inspected
their work and communicated the changes needed to meet the plan.*

*Because of Noah's leadership of the construction project team, the ark
withstood the load of animals and survived the great flood. Noah could not
have built the ark by himself; he needed an involved team to build an ark
that would survive.*

"I believe that everyone chooses how to approach life. If you're proactive, you focus on preparing. If you're reactive, you end up focusing on repairing."

~ John Maxwell

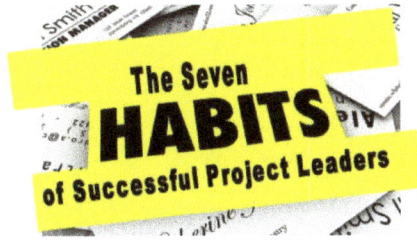

The Seven
HABITS
of Successful Project Leaders

#3 Proactively and Fearlessly Manage Project and Resource Change

As a habit, successful Project Leaders control project and resource change by formalizing a carefully crafted change management process. This process should allow the project team to be empowered with a tool that applies a consistent rubric against every change no matter how small or simple it may seem. This kind of process ensures that every change is evaluated against the project objectives as a whole.

❑ The most successful project leaders understand the value of being proactive and fearless throughout the change management process.

❑ Why does being proactive set you apart from other project leaders? Although many will not admit it, most project leaders think *reactively* and not *proactively*.

❑ Being a proactive leader means that instead of merely reacting to potential project changes as they happen, you consciously anticipate change-related events based on your intuition, calculation and experience.

❑ Being proactive gives you the time and the opportunity to think through the options available and actions necessary to minimize project impact and risks.

❑ Now, when you combine being proactive with the courage…and lack of fear… to *"do the right thing…when it needs to be done,"* you will elevate your leadership presence, your influence and your team's level of success.

From a Brighter Perspective

Proactive Change Management Separates Project Leaders from Project Managers – *Leaders realize that they will face change opportunities every day.*

"No thanks...this project is already behind schedule."

Think of change as part of your day – "What change(s) will I need to consider or make today? "Look for opportunities to make subtle changes to provide momentum.

Resource changes are inevitable – people are suddenly unavailable; external resources are delayed; funding is slower than promised or priorities shift without warning. You, the leader, must proactively address the issue or issues head on.

The team is watching you.

31

"Business people need to understand the psychology of risk more than the mathematics of risk."

~ Paul Gibbons

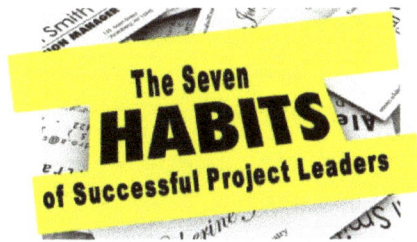

The Seven
HABITS
of Successful Project Leaders

#4 Link Risk to Common Objectives with a Compelling Vision

Most dictionaries define vision as the *"power of discerning future conditions; shrewdness in planning and foresight."* To some extent, we all have such power. Some of us are more inherently capable than others to look into the future…whether it is next year or next week…and plan for specific outcomes. However, the primary and ultimate challenge for most Project Management Professionals lies in how to articulate a compelling project vision. A vision that will "pave the road" that his or her project team will take to achieve the ultimate goal. A vision that successfully links *project risk* to the *common objectives.*

❑ When a project leader confidently articulates a clear and compelling vision, he or she provides the project team members with the tools and knowledge to confidently move forward, execute with clarity and have a strong sense of the *"risk versus reward"* associated with the mission at hand.

❑ The visioning process will also help you construct a leadership action plan to help you guide your team successfully through each project phase and milestone.

❑ At the start of all major projects, most successful project leaders make it a habit to construct a simple table linking or mapping all project objectives to potential risks and the controls that should be established to mitigate those risks …and make the road smoother.

❑ By clearly linking objectives, risks and controls, you and your project team will become more effective in gathering and maintaining documented evidence of the project's status and potential risks. This important "linkage" will also aid in the development of the team commitment and trust needed to achieve the ultimate project goal.

From a Brighter Perspective

A Compelling Vision Guides You and Your Team to Achieving Your Goals

"It looks dangerous. Tell me again, why are we going down there?"

If you were walking across an open meadow and saw a pot of gold next to the end of a rainbow, would you let obstacles prevent you from trying to reach it?

No. You would not.

Just the opposite…you would assess the situation, determine the difficulty and find a way to grab the gold. Project leadership involves showing people the pot of gold at the end of the rainbow and developing a credible plan to take the entire team there.

"One of the true tests of leadership is the ability to recognize a problem before it becomes an emergency."

~ Arnold Glasgow

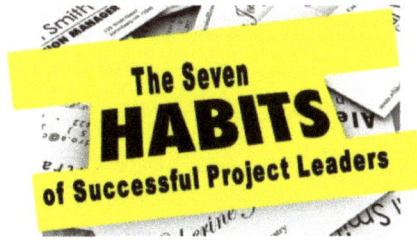

#5 Approach Problem Solving as a Creative and Analytical Process

As the Project Leader, you are not expected to have all the expertise to solve the multitude of challenges and problems that will inevitably surface during the execution of a complex project. However, you are expected to gather the right team of experts and to make sure that the best solutions are found for even the most difficult problems. To accomplish this important task, most successful Project Leaders start by ensuring that his or her team approaches the solution of major problems as both a creative and analytical process...and here's why.

❑ Approaching a problem in an analytical and logical manner is most common, but it normally leads to only a few or the most obvious solutions.

❑ Strategically attacking the problem solving process with more creative thinking usually leads to an array of possible solutions.

❑ Creative thinking starts from the description of the problem and diverges to the identification and evaluation of all available and feasible solutions.

❑ A creative problem solving process requires all of your team's imagination and what many call "thinking out of the box."

❑ Although the two ways of approaching a problem are different, they are linked because one complements the other. This is evident in the fact that all of the solutions that may result from the creative process must later be analyzed to determine an optimum solution based on the current set of circumstances and resources.

❑ Successful Project Leaders are open to assistance from the team in problem solving. They encourage brainstorming.

❑ Successful Project Leaders are not afraid to involve key stakeholders. They will ask, "Which of these solutions do you think is best for you and your organization?"

❑ Successful Project Leaders do not fall back into comfort zones and "cookie cutter solutions"; instead, they seek and develop solutions that fit the project and yield the best results.

❑ As a habit, successful Project Leaders take a creative and analytical approach to problem solving in order to generate better solutions.

From a Brighter Perspective

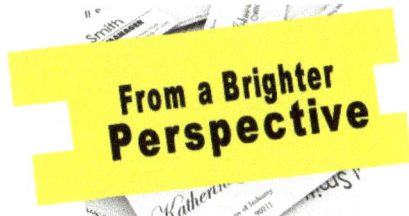

Problem Solving Should Require the Use of Both Sides of the Brain

Math Class
$1 + 2 = 3$
$2 + 2 = Tutu$

"Sorry. The right side of my brain is dominating today."

In the movie "Dead Poet's Society", John Keating, portrayed by Robin Williams, has his students stand on top of the desk to get a different view.

Each young man sees the classroom from a new perspective.

As the Project Leader, you need to do this. View a problem or situation from 360 degrees. Visualize it from the perspective of the client/sponsor, the end user, a third party and as the worst possible critic.

When you do this, you will see different solutions - some better than others.

Work through them with your team and you will arrive at the best solution.

39

"Quality is never an accident. It is always the result of intelligent effort."

~ John Ruskin

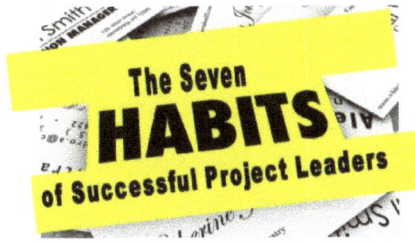

The Seven HABITS of Successful Project Leaders

#6 Maintain Project Quality with Incremental Measures and Team Focus

One of the most difficult tasks of any project is measuring and managing project quality. Most successful Project Leaders faithfully hold to the adage, *"If you can't measure it, you can't manage it."* They also realize that when they employ incremental measures throughout the project lifecycle and maintain a team-wide focus on building quality into their project performance as well as into their products and services, they get the best results.

❑ Quality is the totality of characteristics and features of a product, service or performance that affects its ability to satisfy the stated or inferred needs.

❑ Of course, measuring quality performance and quality outcomes can be achieved with various techniques: including milestones, weighted steps, value of work done, physical percent complete, earned value and other measures. You should carefully choose which technique is the *best fit* for the project.

41

❏ Project quality and performance can be tracked by any appropriate measure – cost, hours, quantities, schedule, percent complete and other measures.

❏ Strategically establishing incremental milestones will allow you and your team to more closely monitor both project performance and quality. Early detection of performance issues could minimize the need for major corrective actions later.

❏ Most successful Project Leaders know that it is difficult to effectively measure the quality of results during the project execution. Therefore, they make it a habit to keep their team focused on improving the overall quality of project performance – which results in the improved quality of project deliverables.

From a Brighter
Perspective

Leaders Keep Their Team Focused on All Project Goals as well as All Incremental Milestones

"Sure glad the hole is not on our end."

Milestones are not millstones – they serve as measures of success, not burdens. Part of your role as the chief communicator and confidence builder is to keep the team (and client/sponsor) apprised of the project's progress.

Each milestone should be acknowledged and celebrated – even if the celebration is only a "nice job people, we completed that phase on time and on budget. Now, on the next phase…"

"Optimism is the faith that leads to achievement. Nothing can be done without hope and confidence."

~ Helen Keller

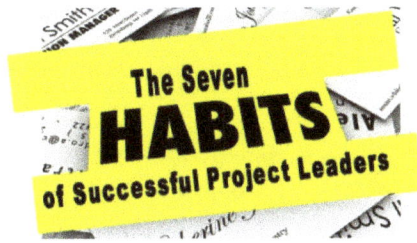
The Seven HABITS of Successful Project Leaders

#7 Accept Your Role as the Chief Confidence Builder

A project team may be comprised of staff members from the same department, multiple departments or even several different organizations. The most common set of characteristics of a good team member will generally include being skilled, experienced, dependable and motivated. However, during the project execution, what distinguishes "good" team members from *outstanding* team members is the level of confidence they have in themselves, in their ability to get the resources they need and in you...the Project Leader.

❑ As a leader first and a manager second, one of your most important roles as a project management professional is to cultivate and build your team's confidence.

❑ Your team members should have confidence in themselves as well as confidence in your leadership skills. In general, team members will judge the probability of future success based on past performance.

❑ As you work with your team through project execution, you will build a track record based on project successes and failures. If you maintain a good record of accomplishment, you will create a sense of optimism that future project challenges will also be successfully conquered. If you have a record of disappointment, your team may lose confidence in you and the probability of the project successfully achieving its goals.

❑ There are many tested principles that you can deploy to help build confidence in your team members. They include holding team members accountable for their actions, providing decision-making opportunities, delegating important tasks and providing stretch assignments.

❑ People respond to "this is what we will do" better than, "maybe this will work". President John F. Kennedy, did not say, "Let's try our best to put a man on the moon." Instead he said; *"We choose to go to the Moon in this decade and do the other things, not because they are easy, but because they are hard; because that goal will serve to organize and measure the best of our energies and skills, because that challenge is one that we are willing to accept, one we are unwilling to postpone, and one we intend to win."*

❑ Taking the time to know your team *as individuals* and *as a group* will help you decide what confidence-building actions are appropriate and when to deploy them.

❑ As a habit, the most successful Project Leaders accept and understand their important role as the Chief Confidence Builder.

From a Brighter Perspective

A Positive Approach Creates an Atmosphere of Success - One of your roles as a leader is to create an environment for success.

"Believe me fellows; all of you are valued team members."

Thomas Edison tried numerous filaments before he discovered that carbon was the best filament. Months after he received his patent, Edison determined that carbonized bamboo lasted over 1200 hours and he switched to that filament.

When asked how he withstood so many failures, Edison replied, "I have not failed. I've just found 10,000 ways that won't work." Edison took what the world considered failure and motivated his team and himself to improve the light bulb and effect a change on the entire world.

"Checklists remind us of the minimum necessary steps and make them explicit. They not only offer the possibility of verification but also instill a kind of discipline of higher performance."

~ Atul Gawande

PART TWO

We mostly use checklists to compile our grocery needs and make sure that we don't arrive back at home having forgotten the milk or eggs. However, checklists, as simple as they are, are some of the most powerful tools available to us…in business and in life.

As we all know, effective project leadership starts during the project conception phase and continues throughout the project's existence. However, it is important to remember that "project management" is not that same as "project leadership."

Most seasoned project management professionals are experts when it comes to applying processes, methods, knowledge, skills and experience to achieve the project objectives. Then again, *leadership* is not a process. Effective project leadership is the strategic use of a combination of interpersonal skills, insights, personal attributes and timely actions, which motivates and inspires others to reach a common vision.

Taking correct actions at the correct time is critical when it comes to building the *winning project relationships* required to strengthen your leadership presence and influence. As a side note, if you are not already doing so, in addition to your project team members, you should also include interactions with your peers, upper management and all project stakeholders on your list of winning relationships.

The *Project Leadership Checklist* is comprehensive and designed to become your personal, experience-based and PMI compatible leadership guide and companion.

The *Leadership Checklist* strategically focuses on the essential elements of successful project leadership. This uniquely crafted and intuitive guide will help you recognize the actions you should take, and when you should take them, during each phase of the project.

The *Leadership Checklist* consists of five sets of coordinated lists. Each set corresponds with one of the five basic phases of a standard project. Within each list, you will find a specific set of actions linked to the key attributes of the *Seven Habits of Successful Project Leaders* as presented in Part One ----*Communications, Problem Solving, Quality Control, Risk Analysis, Change Management* and *Confidence Building*. Since most projects are unique, transient endeavors, undertaken to achieve specific outputs, outcomes or benefits, the *Leadership Checklist* allows you to easily include and track additional actions.

The *Leadership Checklist* is not a "To Do" list. It is a "To Think" List. All of the *affirmations* listed are experienced-based leadership *actions* selected to stimulate thought and serve as a personal catalyst. By using this powerful tool as both a catalyst and a strategic guide, you will find yourself swiftly reaching new heights of project leadership success by proactively, systematically and effectively focusing on project-phased leadership actions and affirmations.

Our extensive project management experience has proven that this type of strategic focus will intrinsically increase your influence and leadership presence on any type or size project.

PHASE I

NOTES

Project Conception and Initiation

During this phase, you and the client/project sponsor(s) will identify if the project can realistically be completed and should be a "Go" or "No Go."

Actions to Strengthen Project Leadership Presence and Influence

Communications	✓
• I have evaluated all aspects of the proposed project concept and have determined if I could become passionate about delivering the anticipated results.	
• I have a strategy by which I will communicate clarity of objectives, targets and benefits with each project stakeholder to highlight the merits of the project to their organization.	
• I have identified conversations with key stakeholders to understand the politics and hidden agendas surrounding the project.	
• I understand the cultural and territorial issues of those who are examining the project to determine its "Go" or "No Go."	
• I understand the diversity of cultures and communication styles of the sponsors and stakeholders.	
• I have developed a list of potential team leads I will request for the project and have interjected my intentions and reasoning into discussions regarding project "Go" or "Go."	
• I have determined to what degree I will risk selling the merits of the project, if I sense "the sell" is too much of an uphill battle.	
• *(Additional Actions)*	
• *(Additional Actions)*	

Problem Solving

	✓
• Based on the basic concept and initial approach, I have identified potential problem areas in terms of staffing, execution and resources.	
• I have identified the knowledge experts that would be required to ensure the project's success.	
• I have solicited insights and opinions from knowledge experts at this stage to assist in identifying potential problems, which should be factored into the project's concept design.	
• *(Additional Actions)*	
• *(Additional Actions)*	
• *(Additional Actions)*	
• *(Additional Actions)*	

Quality Control

	✓
• I have discussed with project sponsors their expectations of overall project quality and the controls I will use to meet those expectations.	
• I have determined to what extent I will get my sponsors and relevant stakeholders involved in the quality control process.	
• *(Additional Actions)*	
• *(Additional Actions)*	
• *(Additional Actions)*	
• *(Additional Actions)*	

Risk Analysis

	✓
• I have analyzed the project concept and have identified inherent risks to delivering the anticipated results and project modifications required to mitigate them.	
• I have interjected my risk analysis and mitigation into project concept discussions with project sponsors and stakeholders.	
• *(Additional Actions)*	
• *(Additional Actions)*	

Change Managment

	✓
• I have outlined the process I will use to evaluate project changes and their potential impact on each project stakeholder.	
• I have shared my change management approach with my sponsors and project stakeholders and solicited their feedback.	
• *(Additional Actions)*	
• *(Additional Actions)*	

Confidence Building

	✓
• I have a strategy for building and maintaining a high level of confidence among the members of the project team during each project phase.	
• I have shared my confidence building strategy with my direct supervisor, project sponsors and key project stakeholders.	
• *(Additional Actions)*	
• *(Additional Actions)*	

NOTES

PHASE II

Project Launch

NOTES

Project Launch

During this phase, resources and tasks are distributed and team members are assigned their responsibilities. This is a good time to bring up important project related information.

Actions to Strengthen Project Leadership Presence and Influence

Communications	✓
• I have reviewed the format and frequency of project reporting with my sponsor and project stakeholders and gained their approval.	
• I have made sure that my team members know that they can approach me with informal discussions and reports.	
• I have documented my vision and mission statement and have considered how I will effectively communicate them during the initial project team kick-off.	
• I have strategically determined how to involve my sponsor(s) and key stakeholders in project kick-off activities.	
• I have strategically determined how to communicate project team responsibilities and expectations.	
• I have developed a plan that ensures our project communication is appropriate and delivered to the correct stakeholders and team members.	
• *(Additional Actions)*	
• *(Additional Actions)*	
• *(Additional Actions)*	
• *(Additional Actions)*	
• *(Additional Actions)*	
• *(Additional Actions)*	

Problem Solving

	✓
• I have documented and verbally shared with the project team the problem solving techniques we will use during project planning and execution.	
• I have discussed the project launch with knowledge experts who may be required to address potential problems and invited them to selected project launch activities.	
• I have developed a method to prioritize potential problems and assign responsibilities to team members.	
• *(Additional Actions)*	
• *(Additional Actions)*	
• *(Additional Actions)*	
• *(Additional Actions)*	
• *(Additional Actions)*	

Quality Control

	✓
• I have outlined my quality control vision and communicated it during the project launch activities.	
• I have solicited team member input on how to implement the necessary quality control techniques and measures.	
• *(Additional Actions)*	
• *(Additional Actions)*	
• *(Additional Actions)*	
• *(Additional Actions)*	
• *(Additional Actions)*	

Risk Analysis	✓
• I have discussed risk management and risk tolerance with the project team.	
• I have solicited team member input on how to implement risk analysis and risk management techniques.	
• *(Additional Actions)*	
• *(Additional Actions)*	

Change Managment	✓
• I have developed a change management strategy, which is a "good fit" for this particular project.	
• I have reviewed the change control plan and change management process with project sponsors and project stakeholders to gain their agreement.	
• I have reviewed the project's change management plan with the project team and have addressed any concerns they may have.	
• I have strategically selected a Change Control Board and defined the role of each member.	
• *(Additional Actions)*	
• *(Additional Actions)*	

Confidence Building	✓
• I have re-evaluated my plan for increasing the confidence of the project team through this project phase.	
• *(Additional Actions)*	
• *(Additional Actions)*	

NOTES

PHASE III

Project Definition and Planning

NOTES

Project Definition and Planning

During this phase a project plan, project charter and project scope definitions are developed and documented. You and your team should prioritize the project objectives, develop a budget and schedule. You should also determine the project resources required.

Actions to Strengthen Project Leadership Presence and Influence

Communications	✓
• I have determined how I will provide *good news* and *bad news* updates to the sponsor(s).	
• I know how I will communicate status to each level of my sponsor's organization.	
• I know how I will address the team when we successfully achieve major milestones.	
• I have a communications plan to discuss missed milestones with the team.	
• *(Additional Actions)*	
• *(Additional Actions)*	
• *(Additional Actions)*	
• *(Additional Actions)*	
• *(Additional Actions)*	
• *(Additional Actions)*	
• *(Additional Actions)*	
• *(Additional Actions)*	
• *(Additional Actions)*	

Problem Solving | ✓

• Based on the final project definition, I have identified potential problems in staffing, execution and resources.	
• I have developed an initial plan to help guide the project team through each problem area, if one or all of the potential problems should occur.	
• I have decided how I will use my functional relationships and alliances to aid in proper and timely problem resolution.	
• I have a plan to involve knowledge experts to help solve complex problems without alienating the team.	
• I have created a "problem solving environment" which allows "out of the box solutions" to be considered.	
• *(Additional Actions)*	
• *(Additional Actions)*	
• *(Additional Actions)*	

Quality Control | ✓

• I have reviewed the quality control program to ensure that it is thoroughly documented.	
• I have ensured that the entire project planning process includes adequate quality checkpoints.	
• I have planned how to communicate quality control results and meaning to sponsors and stakeholders.	
• *(Additional Actions)*	
• *(Additional Actions)*	
• *(Additional Actions)*	
• *(Additional Actions)*	

Risk Analysis | ✓ |

• I have developed a plan to consult with a mix of people, with different perspectives, backgrounds and knowledge to predict, assess and manage risk on this particular project.	
• I have reviewed the project plans and tactical markers with my team members to help anticipate and neutralize potential risks.	
• I have developed a plan to assemble the team to gain their input on potential risks and ways to handle them.	
• *(Additional Actions)*	
• *(Additional Actions)*	

Change Managment | ✓ |

• I have decided how to integrate change management and project management activities.	
• I have a plan to ensure the flow of relevant information during the entire change management process.	
• I have created an environment that allows creativity and positive changes that would benefit the project and the client.	
• *(Additional Actions)*	
• *(Additional Actions)*	

Confidence Building | ✓ |

• I have adequately planned how I will delegate responsibility and accountability to my team.	
• I am committed to including members of my team in decision-making process.	
• I have re-evaluated my plan for increasing the confidence of the project team through this project phase.	
• *(Additional Actions)*	

NOTES

PHASE IV

Project Performance and Control

NOTES

Project Performance and Control

During this phase, you compare project status and progress to the actual plan, as the project team performs the scheduled work. You may need to adjust schedules or do what is necessary to keep the project on track.

Actions to Strengthen Project Leadership Presence and Influence

Communications	✓
• I have scheduled regular communication meetings with my team.	
• I have scheduled regular communication meetings with my sponsor(s) and relevant stakeholders.	
• I make it a habit to review my written communications to make sure that messages are always clear and concise.	
• *(Additional Actions)*	
• *(Additional Actions)*	
• *(Additional Actions)*	
• *(Additional Actions)*	
• *(Additional Actions)*	
• *(Additional Actions)*	
• *(Additional Actions)*	

Problem Solving

	✓
• I have assessed the project team's problem solving skills and determined which team member will take the lead in each project area in coordinating the problem solving process.	
• I have developed and maintain a log of all major problems, resolutions and project impacts.	
• I make it a habit to make everyone feel comfortable in expressing his or her concerns and opinions freely during the problem solving process.	
• I make it a habit to foster an open-minded and positive project environment to facilitate a similar approach to problem solving.	
• *(Additional Actions)*	
• *(Additional Actions)*	
• *(Additional Actions)*	

Quality Control

	✓
• I have established periodic quality review sessions to ensure that all quality control findings and remediation are thoroughly documented.	
• I have created an environment that empowers my team to approach me with quality issues and I do listen.	
• *(Additional Actions)*	
• *(Additional Actions)*	
• *(Additional Actions)*	
• *(Additional Actions)*	
• *(Additional Actions)*	

Risk Analysis ✓

• I make it a habit to ask the right questions regarding schedules and budgets to identify inherent risks.	
• I make it a habit to prioritize risks by assessing their probable impact prior to discussing them with the full team.	
• I have empowered my staff to identify risks and bring them to my attention in real time.	
• I have a "go-to-person" on the client/sponsor's staff to discuss project risk concerns and obtain insight on solutions that will satisfy the client/sponsor.	
• *(Additional Actions)*	
• *(Additional Actions)*	
• *(Additional Actions)*	
• *(Additional Actions)*	

Change Managment ✓

• I proactively manage conflict during the change management process.	
• I make it a habit to focus on the benefits as well as challenges associated with all changes and reinforce them.	
• I make it a habit to listen to objections to proposed changes and address them fully.	
• *(Additional Actions)*	
• *(Additional Actions)*	
• *(Additional Actions)*	

Confidence Building

	✓
• I have developed a method to hold each team member accountable for his/her actions.	
• I make it a habit to encourage each team member to get involved (as appropriate) in the decision making process.	
• I have a plan to delegate decision-making to team leads.	
• I have identified the project areas where I will delegate more responsibility, as appropriate, to help build the confidence of my team leaders.	
• I make it a habit to regularly acknowledge carefully assessed risk-taking.	
• I make it a habit to look for "high potential" team members and encourage them to do more.	
• I have evaluated my team and have identified which team members I should provide "stretch tasks""to help build their confidence.	
• I have evaluated my team and have identified the best way to motivate them and build confidence.	
• (Additional Actions)	
• (Additional Actions)	
• (Additional Actions)	
• (Additional Actions)	
• (Additional Actions)	
• (Additional Actions)	

PHASE V

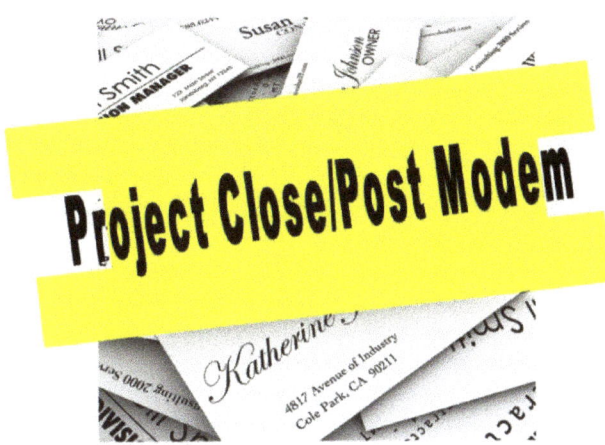

Project Close/Post Modem

NOTES

Project Close/Post Modem

After project tasks are completed and the client/sponsor has approved the outcome, an evaluation is necessary to highlight project success and/or learn from project history.

Actions to Strengthen Project Leadership Presence and Influence

Communications	✓
• I have developed a plan to document the project successes and missed opportunities as they occur for project close and post modem discussions.	
• I have discussed project close expectations with my sponsors and stakeholders to aid in directing my team during this project phase.	
• I have communicated the project close process with the team prior to project close to capture important input from team members before they roll-off of the project.	
• *(Additional Actions)*	
• *(Additional Actions)*	
• *(Additional Actions)*	
• *(Additional Actions)*	
• *(Additional Actions)*	
• *(Additional Actions)*	
• *(Additional Actions)*	
• *(Additional Actions)*	

Problem Solving | ✓ |

• I have developed and shared with my team a *Problem Solving Lessons Learned Review Template* for documenting lessons learned for post modem reviews.	
• (Additional Actions)	
• (Additional Actions)	
• (Additional Actions)	
• (Additional Actions)	

Quality Control | ✓ |

• I have developed and shared with my team a *Quality Control Lessons Learned Review Template* for documenting lessons learned for post modem reviews.	
• I have personally reviewed the final administrative closure process to achieve all sponsor and stakeholder expectations.	
• I have personally reviewed the final contract closure process to achieve all sponsor and stakeholder expectations.	
• I have asked the team to identify and document post-implementation concerns to aid in monitoring quality issues during the initial deployment of the services/products delivered.	
• (Additional Actions)	
• (Additional Actions)	
• (Additional Actions)	
• (Additional Actions)	

Risk Analysis | ✓ |

	✓
• I have developed and shared with my team a *Risk Management Lessons Learned Review Template* for documenting lessons learned for post modem reviews.	
• I have asked the team to identify activities and document post-implementation activities during project closeout to minimize operational risk during the initial deployment of the services/products delivered.	
• I have ensured that the team maintains a record of major project risks encountered, the mediation taken and their impact on the project for future post modem reviews.	
• *(Additional Actions)*	
• *(Additional Actions)*	
• *(Additional Actions)*	

Change Managment | ✓ |

	✓
• I have developed and shared with my team a *Change Management Lessons Learned Review Template* to use to gather lessons learned input from team members for post modem reviews.	
• I have ensured that all approved changes to the baseline schedule have been identified and their impact on the project documented for post modem review in the future.	
• I have ensured that all changes to the cost baseline have been identified and their impact documented for future post modem reviews.	
• I have ensured that all changes to the initial schedule baseline have been identified and their impact documented for future post modem reviews.	
• *(Additional Actions)*	
• *(Additional Actions)*	

Confidence Building ✓

	✓
• I have assured my sponsor and relevant stakeholders that the closeout process and documentation will be thorough, complete and available for future post modem reviews.	
• I have met with my project team prior to project closeout and expressed my confidence in their ability to properly conduct all project closeout activities in a thorough and complete manor.	
• *(Additional Actions)*	
• *(Additional Actions)*	
• *(Additional Actions)*	
• *(Additional Actions)*	
• *(Additional Actions)*	
• *(Additional Actions)*	
• *(Additional Actions)*	

NOTES

"Always carry a notebook. And I mean always. The short-term memory only retains information for three minutes; unless it is committed to paper you can lose an idea forever."

~ Will Self

PART THREE

Project Leadership Diary

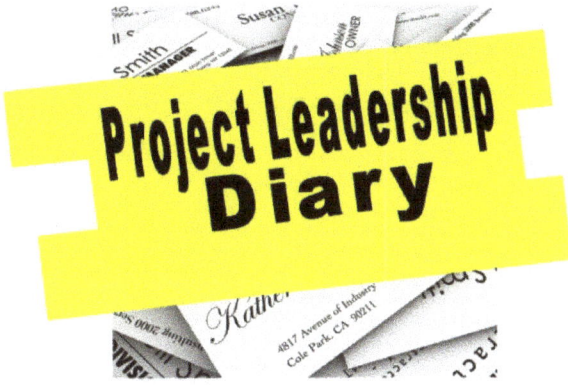

Writing can do wonders for both your personal and professional health.

Successful project leaders can benefit greatly from keeping a *reflective journal*. In particular, keeping a private journal for each project can help keep the proper focus on what you should be doing as a *leader*, during each project phase. It can help define your personal/professional objectives and serve as a roadmap for tracking performance and achievements throughout the project.

The following *Leadership Diary* has been strategically crafted to be *your* reflective journal. It is a unique complement to the *Leadership Checklist*. Your thoughtful responses to each of the six questions during each of the five project phases will help you keep focus on the project's leadership priorities, assess your ongoing leadership performance and enhance your overall project leadership planning.

Tracking your leadership successes will build self-confidence, which is a great contagion for your team. The team feeds off your confidence as the project leader. You will also be empowered by feeding off your team's enthusiasm, trust and willingness to follow your lead.

After you complete each project phase, remember to answer the last question, *"What were my results and what would I do differently next time?"* This is an important professional growth step – applying what you learned to improve in the future.

At the end of the project, you will own a "private reflective journal" to reference later for further development and/or share with other project leaders to help them enhance and grow their professional performance.

It has been our experience that documented leadership *lessons learned* provide the most insight, information and inspiration.

Leadership Diary

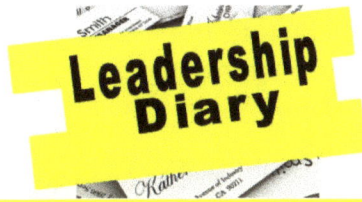

Project Conception and Initiation

❑ What will I prioritize during this *Project Phase* to maximize my leadership presence and influence?

_____

_____

_____

_____

_____

_____

❑ What will be my primary objective(s)?

_____

_____

_____

_____

_____

_____

❑ How will I achieve my primary objective(s)?

_____

_____

_____

_____

_____

_____

_____

❑ What will I need and whom do I need to get involved to ensure I will be successful?

_____

_____

_____

_____

_____

❑ How will I measure the results?

_____

_____

_____

_____

_____

❑ What were my results and what would I do differently next
   time?

_____

_____

_____

_____

_____

_____

_____

_____

REFLECTIONS

Leadership Diary

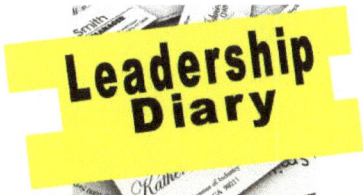

Project Launch

❑ What will I prioritize during this *Project Phase* to maximize my leadership presence and influence?

_____

_____

_____

_____

_____

_____

❑ What will be my primary objective(s)?

_____

_____

_____

_____

_____

_____

❑ How will I achieve my primary objective(s)?

_____

_____

_____

_____

_____

_____

_____

_____

❑ What will I need and whom do I need to get involved to ensure I will be successful?

_____

_____

_____

_____

_____

_____

_____

❏ How will I measure the results?

_____

_____

_____

_____

_____

❏ What were my results and what would I do differently next time?

_____

_____

_____

_____

_____

_____

_____

_____

_____

REFLECTIONS

Leadership Diary

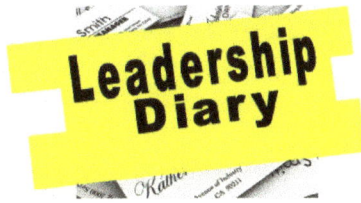

Project Definition and Planning

❑ What will I prioritize during this *Project Phase* to maximize my leadership presence and influence?

_____

_____

_____

_____

_____

❑ What will be my primary objective(s)?

_____

_____

_____

_____

_____

❑ How will I achieve my primary objective(s)?

_____

_____

_____

_____

_____

_____

_____

❑ What will I need and whom do I need to get involved to ensure I will be successful?

_____

_____

_____

_____

_____

_____

❑ How will I measure the results?

_____

_____

_____

_____

_____

❑ What were my results and what would I do differently next time?

_____

_____

_____

_____

_____

_____

_____

_____

REFLECTIONS

Leadership Diary

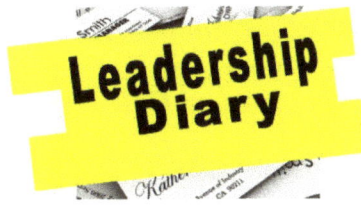

Project Performance and Control

❑   What will I prioritize during this *Project Phase* to maximize my leadership presence and influence?

_____

_____

_____

_____

_____

_____

❑   What will be my primary objective(s)?

_____

_____

_____

_____

_____

_____

❑ How will I achieve my primary objective(s)?

_____

_____

_____

_____

_____

_____

_____

❑ What will I need and whom do I need to get involved to ensure I will be successful?

_____

_____

_____

_____

_____

_____

❑ How will I measure the results?

_____

_____

_____

_____

_____

❑ What were my results and what would I do differently next
time?

_____

_____

_____

_____

_____

_____

_____

_____

REFLECTIONS

Leadership Diary

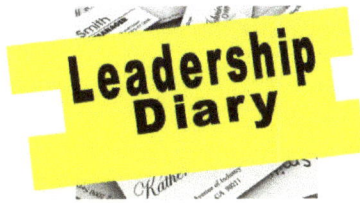

Project Close/Post Modem

❑ What will I prioritize during this *Project Phase* to maximize my leadership presence and influence?

_____

_____

_____

_____

_____

_____

❑ What will be my primary objective(s)?

_____

_____

_____

_____

_____

_____

❑ How will I achieve my primary objective(s)?

_____

_____

_____

_____

_____

_____

_____

❑ What will I need and whom do I need to get involved to ensure I will be successful?

_____

_____

_____

_____

_____

_____

❑ How will I measure the results?

_____

_____

_____

_____

_____

_____

❑ What were my results and what would I do differently next
time?

_____

_____

_____

_____

_____

_____

_____

_____

REFLECTIONS

DRIVING ULTIMATE PROJECT PERFORMANCE™

What kind of magic does the Walt Disney Company use to keep its large and sprawling staff of smiley, friendly, and competent workers all on the same page…and keep them all smiling? Contrary to popular belief, it is not Disney pixie dust. What is actually responsible for the success is a robust leadership training program.

Yes. Good training delivers good results.

Jim Grigsby and Ervin (Earl) Cobb have teamed up to develop and deliver a powerful, new project leadership development program titled, **Driving Ultimate Project Performance™**.

This intensive program is uniquely designed and delivered to help build a personalized *Leadership Improvement Platform* that successful Project Management Professionals (PMPs), Project Managers and Business Managers can utilize to consistently achieve ultimate project performance. The program includes follow-up *one-on-one* coaching sessions and is limited to small groups.

Driving Ultimate Project Performance™ has been described by many as one of the most distinctive, demanding and career changing programs available today.

Contact them today to find out when the next program is being offered in your area. You can also arrange to schedule a custom program at your location for your team or organization.

Ervin (Earl) Cobb	**Jim Grigsby**
earl@richerlifellc.com	Jgrigsby@jimgrigsbyconsulting.com
602.708.4268 (USA)	772.539.1990 (USA)

OTHER LEADERSHIP BOOKS BY ERVIN (EARL) COBB

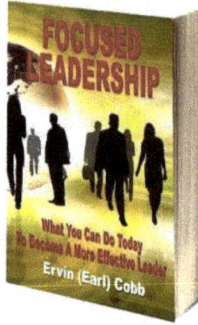

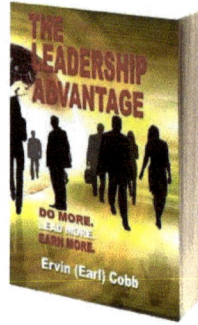

FOCUSED LEADERSHIP

WHAT YOU CAN DO TODAY TO BECOME
A MORE EFFECTIVE LEADER

ISBN: 978-0-9863544-6-5

THE LEADERSHIP ADVANTAGE

DO MORE. LEAD MORE. EARN MORE

ISBN: 978-0-9903291-3-8

SELF-IMPROVEMENT BOOKS BY JIM GRIGSBY

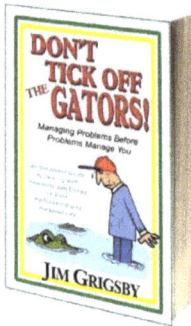

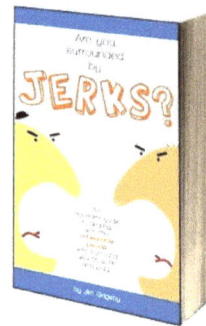

DON'T TICK OFF THE GATORS!

MANAGING PROBLEMS BEFORE
PROBLEMS MANAGE YOU

ISBN: 978-1-5682510-6-6

STORIES THAT CAME TO MIND

FOUR SHORT STORIES
FOR FUN READING

ISBN: 978-1-4524905-5-7

ARE YOU SURROUNDED BY JERKS?

HOW TO DEAL WITH THE
MOST ANNOYING PEOPLE

ISBN: 978-1-4524035-8-8

About The Authors

Ervin (Earl) Cobb, MSEE

Earl Cobb is a Project Manager and a retired technology executive. Earl is an accomplished corporate manager, entrepreneur, author and lecturer on leadership development. Thirty years of aggressively climbing the corporate ladder of Fortune 500, Mid-Market and Venture companies has afforded Earl the opportunities required to accumulate trailblazing leadership experience, insights and success. Earl's ability to reach deep within and his willingness to share his lessons learned with others is what makes his coaching and books unique treasures.

Jim Grigsby, CRCE, CHCS

Jim Grigsby is a Project Manager and the President and CEO of Jim Grigsby Consulting. Jim provides strategic and long-term revenue cycle solutions through analysis, training and engaging inter-departmental collaboration. Jim is also a nationally sought speaker, trainer and lecturer on self-improvement and crisis management strategies, project management, management communications, professional development and mentoring.

╫RICHER Press
An Imprint of Richer Life, LLC

RICHER Press is a full service, specialty Trade Book Publisher whose sole goal is to *shape thoughts and change lives for the better.* All of the books, eBook and digital media we publish, distribute and market embrace our commitment to help maximize opportunities for personal growth and professional achievement.

To learn more visit
www.richerlifellc.com.

www.ingramcontent.com/pod-product-compliance
Lightning Source LLC
LaVergne TN
LVHW010314070426
835509LV00023B/3470